2011-2012

Science Symposium Handbook

Table of Contents

A Letter to Students

Dear Students,

Since the first day of science class this year, you have known that you will plan, research, experiment, and communicate the results and conclusions of your very own science research project.

Everything you need to understand what is expected of you is here in the handbook. You also have classmates, parents and a teacher to help you.

In the end, it will be quite an accomplishment! You will be a participant in the seventh PDS Science Symposium; following in the footsteps of over 400 projects that have been presented in the past years!

So, let's start!

Laura Graceffa
August 2011

Letter from a former student to students doing this for the first time:

The science symposium is not just a science fair that you have at any school but it is a true learning experience for everyone, even Laura. The science symposium is an experience that at sometimes may be overwhelming but when it is over you feel like you have gone through and accomplished so much. Science symposium is an amazing experience and we are truly so lucky to have it in our school... You have to write a lot and edit each paper what feels like a million times but with the help of friends and Laura you can really push through it and come out with a polished finish project that is definitely something to be really proud of...Go into the science symposium with an open mind and then you will be able to take this expertise as a privilege and you will calmly be able to go through it.

--Julia, 8th grader May 2009

Acknowledgements

First, I would like to thank all the students, who, since 2004 have planned and completed such interesting and inspiring projects.

These students have been very generous in allowing me to use their quotes as prefaces to each section. In addition, students have written advice directly to me throughout the project. This has helped me to tailor the handbook to meet the learning needs of students, and I am very grateful for their help.

Over the past few years, many parents have been supportive, especially behind the scenes. Although the spotlight is rarely on parents in a school, they are a vital part of any student's education, and I value all the support you give your children.

Finally thanks go to George Swain and Mark Schlessman who provided helpful comments on the "first edition."

A Letter to Parents

Dear Parents,

Some of the most gratifying moments during the "symposium season," come for me when parents stop to speak about their child's project. For many parents, the symposium gives them an opportunity to see their son or daughter as the "master of his or her destiny:" The student thinks of a project, gets organized, learns more about the science background than the parent could imagine, and then writes about the work with the gift of a muse.

In other families, though, the symposium does not seem to flow as smoothly. Perhaps the student has a hard time getting started, or does not do the work necessary to understand the science. Maybe the student picks a project to meet the first deadline, but then realizes that he or she really has no affinity for the work. Sometimes the words just do not flow when it is time to write.
Take comfort in the fact that over the past eight years I have seldom had a student who fit solely into one or the other above categories. The Science Symposium is a long-term project, and as such, it has ups and downs. Most students will experience times of pride and times of frustration over the next months, and, as adults who have often worked on long- term projects ourselves, we should help to encourage the students to see that a spectrum of feelings is to be expected.

My concrete and seemingly contradictory advice to parents is #1. It is not your project; it is theirs; and #2. They can't do it without you. Walking the line between these two is the trick. Here are some ideas to help you help your son or daughter.

Talk to them. For anyone about to start to act on a big idea, talking it out helps a lot! The parents' role here is to act more as a teacher or coach—ask questions, and offer advice in the manner with which you would engage a colleague. Saying something like, "Have you considered…" or "How would adding ____ change the outcome," is better than making a judgment.

Include an awareness of Science Symposium deadlines in your family planning. This does not mean plan around them; only the May 4th date is that inflexible. My objective in setting goals and breaking down the work into small chunks is that no one will ever start and finish a section the night before it is due. Family plans may necessitate that sometimes a student work far ahead of a deadline to meet it. They can do this best if they are aware of family plans. Please also note that the research study itself should be completed by the week of March 13th. The remaining time until the Symposium is spent analyzing data and writing the paper.

Proofread your son or daughter's work whenever possible. One of the habits that the Science Symposium work tries to instill is that while first drafts can be turned in at school, rough drafts should not be submitted. When I read a student's work I can best help them learn if the student and I are free to concentrate on the science and the content. Having as many spelling and mechanical errors corrected beforehand is very helpful. That said, some parents find proofreading their son's or daughter's work difficult for a variety of reasons. If proofreading is not possible, reminding your student to use the spelling and grammar check, and to read it him or herself, is a good start.

Please encourage them to ask me for help along the way. This tip is in service to advice #1, above. Sometimes when students are frustrated, the teacher is the last person they want to inform, but I really need to be at the top of the list! Many students, especially the experienced eighth graders, will turn in work early, or show me data as it is being collected, just to make sure they are on the right track. I applaud these students for taking charge! Parents can take comfort in the fact that no student has ever failed to complete a project; even when data do not turn out as predicted, students can explain results in scientifically sound ways.

From the time of the initial assignment, until the Symposium week in May, there will seldom be a night when it is impossible for your son or daughter to do the science homework of working on his or her project. During the Symposium "season," from January to May, I schedule some in-class work time for Symposium work. Many students use this for writing and to have conferences with me. It is very important that students plan to bring to school the work they need to do when we have those workshop days. I scale back other homework assignments as well, to give students time to work on their projects.

A list of the due dates is on the last page of this book. Please look it over. A copy of this list is on my portal site, and I include Science Symposium deadlines on the assignment schedules that I e-mail to students. Snow days do not shift deadlines, unless the actual due date is a snow day, and in this case the work must be turned in the next day that we are in school (even if the student does not have science class that day). I will read and return each section of the paper as quickly as possible. Students have to keep track of the revisions necessary for each section, until it is marked "done." Since you have the handbook in your hands now, you may want to flip through it to get a feeling for what is expected, and the type of guidance the students will have. Of course, many aspects of the symposium are taught in class as well—this is not a "do-it yourself" manual. Students also benefit enormously from their interactions with each other during the Symposium "season." I am asking every student to have a parent sign inside the back cover to show that they have seen the handbook. Feel free to remove this letter to parents for your own reference.

Students are required to have a parent read their proposal and initial it. I want to insure that you realize what the project entails, and that the proposal does not commit you to time or work you are unable to do.

Finally, please mark the calendar for Friday, May 4th! I am looking forward to seeing you all there!

In appreciation for all you do to help your child's education,

Laura

The handbook was very useful. I couldn't have remembered all of that stuff! The quotes were really reassuring because it told us what the past students thought about the science symposium.
-- Logan, 7th grade

Forward

From now until nearly the end of the school year you will be designing and completing a unique and personal science project. This workbook will take you through the steps necessary to clearly communicate your research and its findings to the rest of the 7-8 community.

This is a long term project. Although you will begin before winter break, the final event will be Friday, May 4, 2012, when you will present your data to parents, other teachers and students, and visiting scientists.

Why do we have a Science Symposium at Poughkeepsie Day School?

Science class is not exactly the same as the research work real scientists do. In science class you are learning principles, concepts and vocabulary related to an area of science. This means that you are learning about work that other people have already done.

Another aspect of science class, and one that takes up almost half of all class time, is labs. These are hands-on, meaning you get to use equipment, make observations, record data and draw conclusions. You are asked to think creatively about the process and the data, but the teacher almost always tells you what to do and has a particular objective for what you will learn by the end of the period. The inquiry part of the lab lasts only a period or two.

Scientific research is different. It begins with a question that no one has ever asked before, and ends with data that, through the careful and honest interpretation of the researcher, attempts to answer, usually only in part, that question. You live with the question for a long time-- for a professional scientist this may mean years. For you, it will mean most of seventh or eighth grade.

We have a science symposium at Poughkeepsie Day School because we want you to have the opportunity to come as close as possible to experiencing what real scientific research is like. And, like many other assignments across the curriculum, we are looking for ways to make your education personally meaningful to you. Having choice about what to study is one way to find information that is meaningful to you.

The difference between a science symposium and a science fair
Our science symposium is similar to a science fair in some ways. In both, projects are designed by students. The projects are long term. Student researchers must follow the scientific method. The motivation for doing the work, the results and the conclusions must be clearly communicated to the audience.

However, our science symposium is different from a science fair in some very important ways. In a symposium, every participant communicates original research about their findings about a common topic. This means that participants have background knowledge and that allows them to understand and have a genuine interest in each other's work. Symposia are not competitive. Professional scientists participate in symposia, but not in science fairs. Professional scientists organize and participate in symposia so they can share their discoveries, learn from each other, and improve their understanding of science. By hosting a science symposium, we are closer to having an experience that is an authentic science experience.

...for the Science Symposium is a way to experience and look at the world in a different way...

-- Victoria, 8th grade

The science symposium was a great learning experience. I have a sister in 11th grade and she talks about the thesis. I feel like the science symposium will really help us to do long term projects like the thesis.

--Dominique, 7th grade

I thought that symposium was really great experience because now I am able to do whole projects all by myself. Science symposium will probably help a lot in high school too.

--Anya, 7th grade

An Overview of the Assignment

What will I be doing?

You will be responsible for initiating and completing a science project that has something to do with science. Seventh graders will work on a topic related to this year's theme. Eighth graders who participated in the symposium last year may submit a proposal that is a follow up study to the research they did last year, or to this year's theme, or a another topic of choice. You may choose to do an experimental project or a descriptive project. *Either way, the project must generate numerical data.* Most projects are done by one student. If you have a project that would be better done with a partner, you may submit a proposal asking to work as a team. Students working with a partner will still be doing all of their writing alone.

What is the final product?

You will design a poster display of your research, give a 7-10 minute talk about your project, and write a scientific paper about your findings.

Will I have class time to work on it?

Yes, but most of the work will be done on your own time. You will have regular progress checks at school and your classmates may be asked to review your work. ANY missed deadlines for the Science Symposium will result in a letter home.

Can you help me purchase materials?

Most of the materials should be purchased by your family. If you have a project that requires something sold only to schools, I can help you (although the costs will be your responsibility.)

Can my parents help me?

Yes! You will most likely need their help, at least for shopping. If the project is going to be carried out at home you will need some space. If you need to gather data from particular locations or sources you may need transportation. For these reasons, you *must* ask your parents to read and initial your proposal before you turn it in (more about this later.) You should keep your parents up to date about this project. They can be a valuable resource. As a parent myself, I can assure you that parents are usually logical. You can practice your talk in front of them to make sure that it makes sense.

How Will You Be Evaluated?

This is a long term project with many components. Each component has a purpose and is evaluated by how well it meets the requirements of that purpose.

At the back of this workbook is a rubric that defines the characteristics of successful completion of each part of the project. Both you and I may be scoring your work according to these criteria.

Since you know what the best work must contain, you should always self edit your work, using the rubric, before turning it in.

Deciding on a Project

To raise new questions, new possibilities, to regard old problems from a new angle, requires creative imagination and marks real advance in science.

-- Albert Einstein

This is the most important part. When you write this think about how you will record the data numerically. Don't over think and make sure you pick a project that will be fun for you.

--Jesse, 7th grade

Start writing them early so you don't need to rush at the end. Think about what you're doing that year and other subjects related to it that interest you.

--Arthur, 7th grade

Make sure the project fits your personality and try to come up with something of your own. Don't use a friend's old idea or an idea that you overheard in the hallway.

--Dan, 7th grade

The three research questions help you to expand your learning. You can take the three questions and make them into one if you choose.

--Mia, 7th grade

This section takes longer than expected. Although you are only writing three questions you have to think up your whole project for this part. Make sure your experiment is relevant to the outside world. Also make sure that you have an experiment that ties into a question. Otherwise you'll be stuck. But don't sweat it once you have thought of your experiment this section is a breeze.

--Mac, 7th grade

The easiest way to come up with a question is to reflect on your knowledge, or in a more simple tense, what do you want to learn about the world?

--Amelia, 7th grade

I found that you have to be creative when thinking of your research questions, and do not get stuck on one idea.

--Anya, 7th grade

It's not that I'm so smart, it's just that I stay with problems longer.

-- Albert Einstein

Writing Research Questions

Your first task is to write three research questions, one of which you will develop into your symposium project.

Which question you ask will determine how you design a way to answer that question. After your research is finished and you have your data, the original question will determine how you interpret your data.

The other reason why the question is important is that you will be living with it in your mind, and very likely in your house, for months. If you chose a question that does not interest you strongly, perhaps just to get the initial assignment done, you may not be pleased in the upcoming months when you are working and writing about it. The students quoted on the previous page thought that this was the most important advice to pass on to students about to begin a symposium project.

Types of projects

Just like real professional scientific studies, symposium projects generally fall into two types, *descriptive* and *experimental*. Both types generate numerical data — and this is a requirement of the symposium assignment.

In a descriptive study, an organism, environment, or a set of samples are measured in some way. Often the researcher is interested in comparing them, and expects and predicts that they will differ. The researcher, however, does not attempt to change what he or she is studying — the organism, environment or set of samples, is *described* just as it naturally occurs.

In our class work, we are doing descriptive work on water characteristics. We are describing the soil samples and comparing them to each other's. For this part of our study, we are not changing the water in any way.

Descriptive studies will often fit into a title like, "A comparison of _____ and _____," where the names of two or more environments, organisms or samples fill in the blanks.

In an experimental study, the researcher knowingly changes some part of an environment — almost always a controlled environment — and compares this to another environment where no change has been made. Many times these experiments are done on some kind of organism: the researcher then interprets the data to try and predict what would happen to living things if that change were made in the natural environment to natural organisms.

In our class work, we will leaf pack content or location and place them in the stream. We will compare the macroinvertebrates in leaf packs to see if the conditions change the population of macroinvertebrates in the packs.

An experimental study can often fit into the title, "The effect(s) of _____on _____," where the first blank is filled in by what the researcher is changing, and the second is filled in with the name of the organism or environment.

Both types of studies are valuable and you should feel free to design a study of either type.

Feasibility and Relevance

At the same time that you design your questions, you need to imagine how you would test them.

Submit only questions that can be tested. This means your proposed work is feasible. The world is full of interesting questions, but only by formulating the right questions can you generate data that can be interpreted in such a way to answer the question.

Make sure your question is relevant. The opposite of relevant is irrelevant. An example of an irrelevant question would be, "Do plants prefer to be watered with Coke or Pepsi?" This question is not relevant to the real world since Coke and Pepsi are not dumped on plants. The answer would not tell you anything interesting about plants (or Coke or Pepsi, for that matter.)

Here are some tests to put to your research questions:

Will I have access to the materials I need to do the study?
A study on moon dust would not be feasible, for example. Even collecting river water will not be possible unless you have an adult who is willing to drive you to make the collection.

Do I have enough time and good weather between now and March to collect these data?
You would not be able to study tree growth, for example, between now and March. If you want to collect water from a stream, you will need to get it before the stream freezes.

Can I imagine what I will be measuring?
You should be able to collect the numerical data by measuring length (meter, centimeters or millimeters), mass (grams), volume (liters or milliliters), or by counting. Measuring pH and conductivity also give numerical data.

Can I imagine how I will show the data in tables and graphs?
Although you will not complete your data presentation until later in the spring, you should be able to envision, right from the start, what the skeleton of the data presentation will be like. For example, if you are going to graph the data, what will be on the X and Y axes? If you are going to show the data in a table, what will the column and row headings be?

Who cares?

If you ask a question and cannot think of why anyone would care about the answer, the research question may lack relevance. Adapt the question or start over. On the other hand, do not confuse relevant with practical. You can have a good question that seeks to improve our understanding of the world, while not changing the world.

How will I build in replication?

If you were an alien and visited Earth and took one human back to your planet with you, how close would you come to showing the other aliens what a human is like? What if you took two humans? Four? Eight? Replication is necessary to avoid results that are only true for one individual or treatment. You must plan replication into your study. Plan to sample more than one time. If you are growing plants, this means you will need a small population. If you are testing at a site, you will need to visit it more than once. How you replicate will depend on your study, so you may need to get advice.

One final reminder

When you fill in the worksheet on the following page be sure to think carefully! Remember that one of these will be developed into your project—you will be living with it for a long time!

Scientific Measurement

When you are planning your study, think about the measuring tools you will need!

Remember that all measurements in science are made using the **metric system**. Therefore, you will not be making any measurements in inches, cups, tablespoons, ounces, pounds, or degrees Fahrenheit.

When you are working at home, you may, at first, think it is difficult to measure in metric. However, finding a metric ruler is easy—many rulers measure in both sets of unit. Similarly, many measuring cups have metric volume markings, and even kitchen scales can be set to measure in grams. Many thermometers have both Fahrenheit and Celsius markings.

I think that you should give us one or two examples in the [hand]book.....
--student suggestion

Examples of Research Questions

Does the pH of snow change when locations are compared?

Does ultra violet light affect the growth of phytoplankton?

Is solar energy economically feasible in the Hudson Valley?

Does dissolved oxygen change through the season in a stream?

Using the Internet

Sometimes students have found that browsing the internet for ideas is helpful. There are a number of web sites with science fair project ideas. Just remember, you will need to adapt any project idea you find to make it your own original work!

Research Questions Rubric				
Category	**4**	**3**	**2**	**1**
Research Questions	Wrote three interesting questions that were relevant and feasible to study. Replication is planned. How the experiment will be designed is clear.	Wrote one or two interesting questions that were relevant and feasible to study. Replication is planned for these questions. Methods not clear.	Wrote one interesting question that was relevant and feasible to study. The other questions were not testable or were not relevant. Replication not planned.	Did not write questions, or none of the questions could be studied.

Plans for Plants

Many students choose to grow plants for their project. Plants make great study organisms. If you are planning to grow plants, there are some things you need to consider.

You must grow the plants from seed. Picking something that germinates quickly and easily is best. Beans and marigolds are two nice choices. Growing them in individual peat pots works well, and you can set the pots in a waterproof container and water from the bottom after they germinate.

Every plant must be in its own pot. Otherwise you don't know if the crowded conditions influenced your plant, or if your experimental treatment did.

You have to grow a small population of plants for replication. You can't compare just two individuals. You will need to check with your parents to make sure you have room in your house for 10 or more plants.

Plants need time to grow. Most need a couple of weeks just to germinate, so you will need to plant your study in early January in order to have data by March. Make sure your parents are ok with this and can take you shopping for seeds, pots and soil.

Plants need warmth to germinate and sun to grow. Make sure this is possible at your house. You should reshuffle your plants at least every other day so that they are equally affected by the small differences being nearer to a window or drafts makes. Also, many people have pets that like to eat their plants. You will need to control this somehow. If you are travelling over February vacation, you need to account for this so your plants don't die and so that you have the measurements you need.

Most studies do not require that the amount of water is measured, unless that is what you are studying.

Name_____

Research Questions

Your assignment is to think of three research questions. One of these will become your science symposium project. You can write three questions about the same project if you have a consultation first. Successful students consult while doing this assignment.

Research questions must be feasible. For example, you would not be able to do a project characterizing the density of moon dust, because you have no access to moon dust. They also must be relevant and pass the "who cares?" test.

This assignment, then, requires that you not only think of questions, but also write a sentence or two about what the study would be like, should this become your project. Use the back or attach additional sheets if necessary.

1._____
_____
_____

What will you do to answer this question?

2._____
_____
_____

What will you do to answer this question?

3._____
_____
_____

What will you do to answer this question?

Writing the Proposal

First start by researching. Use many different sources and search lots of different topics related to your study. Make your proposal flawless so it is easy to make your intro. The majority of your proposal should be background research and the last paragraph about what you propose to do as your study and how you will go about it.

-- Sarah, 7th grade

Learn a lot of background research before you start your proposal. It will be much easier for you to write if you learn as much as you can about your topic. Take notes! Do not be one of those students who read information about their topic and then goes straight to writing their proposal. You want to read information, take notes and then write your proposal from the notes not straight from the source.

--Julia, 7th grade

DONT USE THE WORD YOU!!!!!

--Ethan 7th grade

Go to the page in the handbook called "An Organizer for Your Research Proposal" it really helps you to get started, especially when you are having a hard time figuring out what to say.

-- Rosie, 8th grade

The proposal is not something to take lightly… because eventually it will become your introduction, which is one of the most important parts of your project. If you don't know what to write it means that you have not done enough research. Take as many notes as possible beforehand

-- Ava, 7th grade

This is when you introduce your project into the world. Include why your project is relevant and why people would care, like your selling your project to the world.

-- Angela, 7th grade

In the last paragraph, write what you are going to do.

--Naseem, 7th grade

In the proposal you're writing about your topic, in a broad way. So, you need to expand your mind to think about how to make it go on, without getting boring. Writing a great proposal can lead to making an excellent introduction.

--Violet, 7th grade

Writing Your Research Proposal

After your question(s) and general project idea have been approved, you need to write a research proposal.

Proposals are not just limited to science. In the business world people write proposals to get investors for new companies or products. When trying to find new clients, business people will write proposals for the work they plan to do. In the arts world people write proposals to secure investors or grant money for performances and fine arts. In the professional scientific world, a research proposal is usually written to secure funding for a research idea. No matter which field you are writing a proposal for, it is very important that the proposal be well written and clearly thought out. You are making a first impression and asking the reader to trust you, and in most cases, *pay* you to do your work.

Writing the proposal can be demanding because you have to learn about the scientific background of your topic at the same time as you are deciding on the details of how you are actually going to do your study. All the work you do now, though, will result in saved time later. The more you learn now, the easier it will be to write your scientific paper later. The more you plan for your research project now, the easier it will be to carry out the study later.

Your proposal has two parts, an explanation of the reason **why your study is important**, and, a description of **how you are going to do the research**.

Why the study is important

Imagine that the reader is asking, as he or she picks up your proposal, "Why should I care about this person's idea?" In order to make the reader care, you need to relate your research question to some larger topic that affects the environment and/or humans. The first paragraph of your proposal, then, will not be about the experiment or study you plan. It should not be about you, at all. It is about the big issue that led you to want to do an investigation.

Your research question must relate to scientific knowledge in general. The focus of your work will be narrower, and yet you should help the reader of your proposal "connect the dots" to a larger, important issue. You will be learning new vocabulary relating to the background of your topic—you will need to define technical terms, that is, scientific vocabulary not used in everyday speech, for the reader.

To make your case convincing, you must back up your argument by citing sources. Your research starts with understanding the basic science and, in some cases, related studies, about your topic. Eventually the writing you do now will go into the introduction of your final poster and paper.

How you are going to do the research

Addressing this part will be unique to your study, and it will follow the introductory paragraphs, described above, in which you make the reader care about the issue you are planning to research. You will start these paragraphs by describing the project. Exactly what you write in this description, though, depends on your study. You may need to explain the sampling methods for a comparative study, or describe variables in an experimental study. You should also describe the amount of time your study will take. For example, if you are taking samples, how often and over which months will you sample? If you are doing an experiment, what will you measure and how frequently will you make your measurements?

Be as specific as possible. For example, if your study uses plants, you must say how many, which species, and where you are growing them. If you were going to change something about their growing conditions, you would need to describe that also.
If you are sampling soil, where will you collect the soil from? If you are measuring or collecting precipitation, where will your collection site be?

Sometimes students include drawings or photographs as part of their proposals. This is a fine idea if it helps explain your plans. Just be sure to give each one a title, label it as a figure (more about this later, but read the Writing Your Results section) and refer to it in the proposal when you want the reader to refer to (i.e. look at) it.

Format and Syntax of the Proposal

Format and syntax will be important in every part of the writing of the scientific poster and paper. Format refers to the structure and arrangement of the paper as a whole. Syntax refers to how you use words in sentences.

Format: Like every other part of the project from now on, you will submit the proposal typed. It is impossible to give a definite minimum page length, but most students seem to be able to accomplish a strong proposal within two pages of writing. This would mean about two or three long paragraphs explaining why the study is important followed by one to two paragraphs explaining how you are going to do the work.

Because you will be doing background research you need to cite sources! Every proposal will have a literature cited section. The next section of this workbook will help you get started doing background research. The format of the literature cited section is the Modern Languages Association (MLA) format that you have been (or will be) taught in your humanities class.

More information about formatting written work for this project is in Appendix II.

Syntax: When you describe the work you propose to do, you will use the future tense. For example, "I will sample water every week…," or "I will measure height every other day…" Since the proposal has not been accepted yet, the future tense is appropriate.

When you write your scientific poster and paper, you will use the past tense. Look for the sub title *Format and Syntax* in each section to find out more about this.

You may also use the word "I" and by all means, write in the first person—except in the first paragraphs of your proposal and introduction. You are doing the work, and so are allowed to be the subject of some of the sentences!

You will not be writing in the second person for any part of your proposal or scientific paper, so banish the word "you" from this project.

Don't use rhetorical questions in your scientific paper. What is a rhetorical question? Well, you just read one! (It is a question that you ask but don't expect an answer to.)

Parent proofreaders

The proposal is the one part of the project that you **must** ask your parents to read before you hand it in. The parent who reads the proposal must initial the bottom of the first page. (After you write other sections, I advise you to have a parent proof read them.)

The reason why parents read the proposal is because the research will be happening under their roofs! If you are depending on them to supply money, time, transportation or space, they must know about it sooner rather than later.

Terms You Should Know

Variable: The category of what you are changing and measuring or monitoring. When studying the weather, for example, some variables are temperature, pressure, humidity and amount of precipitation. Both descriptive and experimental studies have variables.

Control: When you do an experimental study, you leave one group alone (the control) and change another group (or groups.) By comparing the two, you interpret your data to understand whether or not the treatment had an effect. Descriptive studies can also have controls, of a sort. In this case it can be the condition at the beginning of a period of time, or at a starting location.

Treatment: When you do an experimental study, you "treat" one or more samples, and leave one alone as the control. The treatments are linked to the variable. You should refer to the treatments by name, not number. In a descriptive study, each site or time when samples are taken is like a treatment.

"The Literature": This is where you find your background research. It can include textbooks, encyclopedias, articles, and internet sources.

Proposal Rubric				
Category	**4**	**3**	**2**	**1**
Proposal	Background research shows that the study is relevant and that it relates to scientific knowledge. Technical terms are defined and used correctly. Variables are identified and described, and how the results of the study will be measured is included. Foresight and planning are evident.	Background research is present, but not in depth, or not related to proposed study. Terms may be defined, but not used correctly. Not all variables are described, nor are the measurements explained. More foresight and planning are necessary.	Background research shallow and not clearly linked to proposed study. Variables not described. Measurements not indicated. Poor description of how the work will be done.	No background research. No description of how the work will be done.

One thing that is useful is outlining the process of writing.
--student suggestion

An Organizer for Your Research Proposal

I. To focus your proposal, start by writing a working title for your study (you can change it later.)

II. The first paragraph should be general. DO NOT WRITE about your study in this paragraph. You are establishing relevance by hooking your audience with the BIG picture.
List three areas of science and/or the environment that connect to your project:

1.

2.

3.

Write the topic sentence of your first paragraph below:

III. The second paragraph should narrow the focus to a local or more immediate area. It also makes the why of your study clear. You may need to define some terms here.

List terms that are specific to your study:

What is the "why" or reason for your study? How is it relevant?

IV. The last paragraph(s) answer the what, when, where of your study:

What are you collecting data on?

When are you collecting the data?

Where are you collecting your data (or samples)?

How will your study have replication (p.15)?

A Sample Proposal

Gray water leaves households at an enormous speed everyday. For an average family, two hundred and eighty gallons of gray water is wasted (Lamb, 2011). The water that runs down the drains of bathroom and laundry sinks, washing machines, showers and bathtubs all become gray water (Scholz, 2011). Every time someone brushes teeth and the water is left running, a gallon of water is wasted (Nature's Head, 2011). Around one hundred and twenty gallons exit houses every day (Gray Water Facts).

Gray water can be recycled, but not many people know about benefits of recycling gray water (Gelt, 2003). The gray water that is recycled can be used for landscape, or other household needs. Not many people recycle gray water, but with the right tools it can be easily done (Central Coast Gray Water Alliance, 2011). If gray water is shown to be healthy, then so much water will be conserved, and it will help the planet. If gray water is helpful for plants that are eaten, not only will it conserve water, it will also make the growth cycle of plants speed up.

I will test to determine if gray water acts as fertilizer on plants. I will grow Bush Beans and water with gray water. Another group will be watered with tap water. For replication I will grow twelve plants as control, and twelve plants watered with gray water. To compare growth I will measure the heights of the plants every other day. My hypothesis is that the plants grown with the gray water will grow faster because of the nutrients in the water.

The first paragraph says nothing about the study. It has many, many sources cited. It is organized from big issue to more specific. It establishes clear relevance.

The second paragraph defines important terms, like gray water. It makes a bridge between gray water and growing plants. The project still is not mentioned.

The last paragraph is all about the project. Each sentence is written in the future tense and explains the study. The last sentence has the hypothesis. For some studies, a purpose instead of a hypothesis is better suited.

Background Research
Finding Information in the Literature
Laying the Groundwork for your Literature Cited Section

Sometimes learning background research can be overwhelming, especially when you do a search on Google and you do not know what information to use. Do a quick search on Google and bookmark pages that seem interesting to you. Then go back to the bookmarked pages and decide which websites to spend more time reading. This will prevent you from getting overwhelmed with too many sources at once.

-Julia, 7ᵗʰ grade

This is easy if you have the right words. When you type the words, you will be able to find a lot of research. The keywords are really important when you are reading and finding the background research.
--Dominique, 7ᵗʰ grade

Clearly read each section instead of skipping paragraphs, it will give you a better understanding about your project so you can write an amazing introduction and get a good start on your project.

--Jenna, 7ᵗʰ grade

Just because you can't find any information within the first page of a search doesn't mean it isn't there, so don't be afraid to take a while looking for web sites.

--Aidan, 7ᵗʰ grade

Some sources might not be directly connected to your project but some of them might give you ideas of other things to look up. You might find tools that are used to study you experiment, if you look up some of those you might be able to find out more about your project.
--Maggie, 7ᵗʰ grade

The literature cited was kind of difficult because I forgot what websites I used, and so REMEMBER TO SAVE THE LINKS and put them in your bookmarks…
-- Sam, 7ᵗʰ grade

*Google the [key] words that relate to your project, and **really read** [what the search reveals.]*

-- Marissa, 8ᵗʰ grade

Using Key Words to Perform a Literature Search

In everyday English, literature means great written works. For scientists, "the literature," means research that has been written and published in a peer reviewed journal. Peer reviewed journals are those where research papers are submitted, and then sent out to two or three scientists who do similar research. These reviewers read the papers, edit them, and recommend to the journal whether or not the papers should be published. The process is lengthy, but it helps to insure that the work that is published is relevant and honest. Scientists themselves learn much from the comments and criticisms of their (usually anonymous) peer reviewers, although having to read those comments does not always put the researcher in a good mood!

The published literature relating to your topic may be too technical for you to understand this year. However, the general background information will not be, and neither will the basic scientific principles that make your research project relevant and interesting. If you happen to find similar studies to yours that you can read and understand, consider yourself lucky, and by all means include them in your proposal and introduction.

Students who find literature searches frustrating are often not looking in the correct way. Many times this means that their search has been too **broad** or too **narrow**, or that they have not been willing to **read** what they found. You will not find an entire book or even a whole chapter in a book, which relates directly to your study. Instead, you will be looking for paragraphs, or even sentences that are relevant.

Paragraphs or sentences??!! These small amounts of information are useful to you because you are doing a unique and original project, and because you are writing a few paragraphs of background information, not several pages. But, remember, you must be willing to **read** in order to find these relevant sentences and paragraphs.

Identifying keywords

Professional scientists are so careful about making good use of their time, and so inclined to work efficiently whenever possible, that they often place keywords in a labeled section right at the top of their published paper! The keywords identify technical terms in the paper and help a reader to know if the paper will really be worth his or her time to read.

For example, a paper with the title "The Effects of Road Salt on Marigold Plant Growth Rates," might have listed the keywords: *garden plants*, *rock salt*, *plant growth*, *plant germination*, *winter road maintenance*.

Words in the title are also key, but the keywords list provides more words about the topic of the paper that might not be obvious just from the title. There are clues about who might be interested in reading this—people who study marigolds in particular, people who are interested in rock salt, people who study plant growth and germination, and people who study road maintenance and/or pollution. That is a broad range of people!

Before doing any search, you should list keywords related to your study. As you read, you will add to the list…. because you will be learning the "lingo" of your field of study!

Where to look

The first places to start looking for information about your topic are your class notes and textbooks. Make sure you look at *Environmental Science*, by Karen Arms, as well. You brought this big blue book home the first week of school.

Another place to look is in the library. We may have books at school that are on broader topics but have some general information that you will find helpful. The school subscribes to some data bases that can be useful. Your town library is another place to look.

The internet

Most students get most of their information from web based research. By keeping a few points in mind when using it for your literature search, you can find the best quality information.

<u>Remember that anyone can put information on the internet!</u>
Just because someone wrote it and posted it, doesn't mean the information is accurate or up-to-date.

<u>Not all web sites are equal</u>
Generally speaking, sites with *.edu* are the most reliable. For your project you may be able to find the research site of a professor who does environmental work related to your topic. You might also find lecture notes from college classes related to your work.

The ending *.org* can also be good. Many natural history museums, professional societies, or environmental organizations have web sites. These institutions are almost always not- for-profit and maintain web sites specifically to educate the public.

Similarly, *.gov* sites can also be safe and useful. Agencies often maintain information for the public and post the information on the web.

Be careful with **.com**, the most common web address on the internet. While not necessarily bad information, you are more likely to get a "homegrown" posting that may not be accurate and up-to-date.

<u>Find the source</u>
Someone wrote every word you read on the web! This is the author, and the person you are seeking to acknowledge in your literature cited section.

If you use a search engine, like Google, for example, you will likely be taken to the middle of a web site. You may not find the author here. You will have to chase through the site to find the homepage of the site. This is your best bet for finding the author(s) of the web site.

It is not always possible to find the name of the person who actually wrote the words, although this should always be your goal. Sometimes you will have to cite the organization that sponsors and maintains the web site. If the web page has a "contact us" listing, look there. You might even write to ask how the site would like to be cited.

Whether or not you find the author, you also need to look for the date the web page was created, posted or modified. This is analogous (i.e. similar to) the date a book or magazine article was published—so it is an important piece of information.

Keep track of the sources

You can only state information in your paper if you give attribution, so keep track of your sources. For books and articles, you can copy the necessary information onto a note card, or the paper you take notes on.

For web sites hand copying is usually difficult, time consuming, and, given the length of some URL's, frequently a source of error. A better way to keep track of web sources is to bookmark sites that you visit, even if you are not sure you will be citing that source. Another way is to print out at least one page from the web site. The exact web address should print at the bottom of the page.

Writing Your Literature Cited

Anything you read when doing background research for your proposal can and most likely, must, be cited in your introduction. The format for citing sources in scientific papers is different from the format in other areas of writing.

Scientists do not use direct quotes

In science, explaining concepts is the goal of the writing. The exact word choice is usually not as important, as long as the concept is conveyed clearly. When you read someone else's writing, and then use this information in your own writing, you must be sure that you are giving them credit for the concepts they have written about. When you read for background research you must understand the material well enough to restate it your own words. Scientists do not use direct quotes, **and** they do not copy word for word, **ever**! This is plagiarism and it is unethical and against the law. Rearranging words "just a little" is the same as copying word for word.

Restating a concept in your own words takes several steps, some physical and some mental:

1. Read the information
2. **"Talk to the wall."** Say in words, either in your head or out loud, what that source just taught you — without reading from the source.
3. Write a few bullet points of notes about what you read. These are notes, not full sentences. If you find yourself looking at the source for word choice, you haven't understood what you read well enough yet and need to go back to steps one and two and repeat them.
4. When you write your introduction, use the notes you made, not the source, to compose your sentences, in your own words.

Scientific papers do not contain footnotes

The authors of the sources you use will be identified as part of the sentence(s) you write about their work. The authors will be identified by last name and the year of publication of what they wrote. Parentheses help to keep this information from interrupting the reader's attention from what you wrote. You will list the full citation in the literature cited section of your scientific paper, and the reader can find more information about the publication there.
Here are some examples:

Acid rain is damaging to aquatic ecosystems (Smith 2003).

An alternative format:

According to Smith (2003), acid rain is damaging to aquatic ecosystems.

Or:

Smith (2003) found that acid rain is damaging to aquatic ecosystems.

Notice that the citation belongs to the sentence and must be enclosed by the punctuation of that sentence. Don't let citations dangle in the space between sentences.

Here is an example of the wrong format:

Soil texture helps determine plant root growth. (Jones 2003) For this reason we need to assess soil texture.

The "(Jones 2003)" is hanging in space between the period of the first sentence and the capital letter that starts the second sentence. It must be enfolded by the period of the first sentence:

Soil texture helps determine plant root growth (Jones 2003).   For this reason we need to assess soil texture.

Scientists try very hard to acknowledge everyone
You may find, especially when reading about the broad concepts related to your study, that several of your sources make the same important point. Which one should you cite? ALL of them! And, you should be proud of your thorough and hard work.

Here is an example:

Acid rain is damaging to aquatic ecosystems (Brown 1999, Smith 2003, Jones 2004).

In the literature cited section, Brown, Smith and Jones will each have their own entry. Notice that the sources are cited from oldest publication date to most recent. (In the literature cited section the entries will be listed alphabetically, by author's last name.)

The most important point for you to remember about citations
Students NEVER get in trouble for citing too many sources-- they get in trouble for citing too few, or for copying, or practically copying, from the source.

Format and Syntax

Format: Information about the MLA format for citing books, articles and web sites is in Appendix II. Read this *now*, before you start your research, so that you will be sure to collect the necessary information. This appendix also has forms to fill in to help you keep track of your sources. MLA is not the format most used in science, but, we will make the procedure simple and only ask you to learn one format in middle school.

There are some excellent web sites, like *bibme.org* or *easybib* that will help you build a literature cited section in the correct format.

Justify the margin left (not center), and use a "hanging indent." A hanging indent means that the first line in justified left and the subsequent lines in the entry are indented. This paragraph has been formatted with a hanging indent.

Alphabetize the list by author's last name. If you are using more than one source from the same author, list the oldest date of publication first.

Do not number the list.

Syntax:

<u>Varying sentence format:</u> Use the structures of the sentence examples above when writing about your background information. You would find those formats used in any published scientific paper. Sometimes students worry that this will sound repetitive. Don't worry. First, you have three sentence frameworks you can switch among. Second, sometimes the work or concept cannot be explained in just one sentence. So, you will expand on each citation for a sentence or two, and this will vary the sentence structure also.

Name:_____

Identifying Keywords and Starting an Internet Search

1. One good place to start identifying keywords is with your title! Scientific titles often follow the format "The effects of _____ on _____.If you have not thought of title yet, do that first and write it here:

2. List, in the left column below, five keywords that relate to your study. Leave the right column blank for now.

_____   _____

_____   _____

_____   _____

_____   _____

_____   _____

3. Navigate to Google. For your first search, enter all five words that you listed above for #2. This is the narrowest search. If you don't find any pages, don't worry. Broaden your search by trying the words in smaller combinations. Finally, try each word alone.

4. If you find pages that seem useful, **bookmark** it in your Favorite list. You will go back on your own time to read the web site more carefully. Set up a bookmarks folder for the science symposium and put all your bookmarks there.

5. Have another look at the list of keywords. Now that you have "bumped" into a few sites, use the right hand column of #2 to add two to five more words that relate to your topic.

6. Continue searching and reading. You should find that web sites are becoming more specific to your topic and more helpful.

Writing the Introduction

When creating the introduction, do not freak out and start writing a whole new document, just remember your proposal and use it to help you.

--Kelly 7th grade

First change your proposal from future tense to past. Then go through and see how you can make it better. Add to it and make sure you have your peers edit it before you hand it in. Remember to cite your sources as you go.

--Sarah, 7th grade

In your introduction, it is really important to really make sure that you talk about how your project connects to the rest of the world because during the symposium people asked about that.

--Sophie, 7th grade

When you write your introduction, make sure you have your proposal with you.

-- Will, 7th grade

Get this done early; a first green dot is a real confidence booster.
Parenthetical citations are essential: you can never have too many.
Review it several times over the course of several days before turning it in the first time.

-- Tyler, 7th grade

If you write it well, it means you really understand everything fully.

-- Marissa, 8th grade

The introduction takes time and you do not want to just say, " Oh I should get it over with," you should get it done but make sure you do a good job and make sure your happy with what you wrote.

--Rita, 7th grade

Don't wait 'til the last minute.

--Eli, 7th grade

Writing Your Introduction

The introduction is the first part of your paper that a reader will read, after the title. You have to accomplish three purposes when you write it:

1. Justify to the readers why they should care about your research.
2. Teach the readers the important background information for the subject.
3. State the hypothesis (for an experiment) or purpose (for descriptive work) of your study.

If you wrote your proposal well, you have background information and your very own words to draw from to accomplish this!

Like the other parts of your scientific paper, the introduction will have a very structured format. Following this format is required—and it makes the writing easier.

Funneling the reader's attention – the spotlight's on you – eventually, but not initially!

Think of the introduction as being an inverted triangle, inverted pyramid, or funnel that takes the reader from the broad concepts relating to your work to the specifics of what you did:

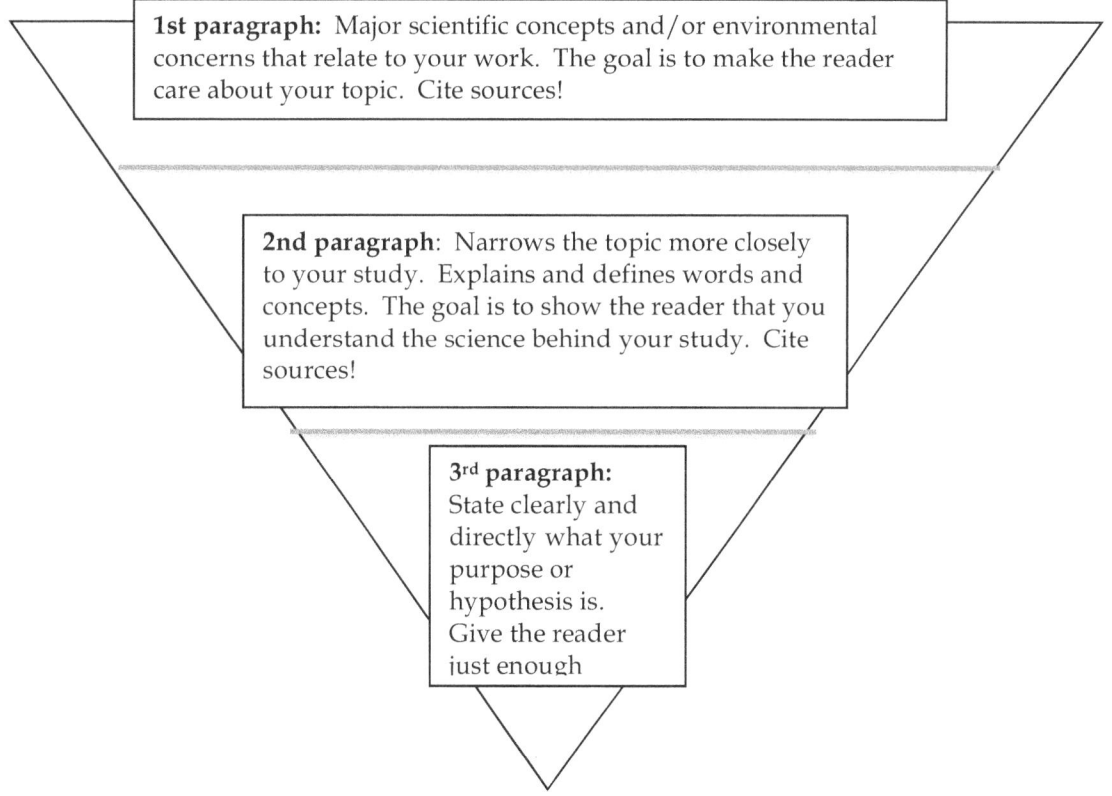

1st paragraph: Major scientific concepts and/or environmental concerns that relate to your work. The goal is to make the reader care about your topic. Cite sources!

2nd paragraph: Narrows the topic more closely to your study. Explains and defines words and concepts. The goal is to show the reader that you understand the science behind your study. Cite sources!

3rd paragraph: State clearly and directly what your purpose or hypothesis is. Give the reader just enough

In your first paragraph, you will *not* mention *your* work at all. You will teach the reader to care about the topic. By the middle of the introduction you will teach the reader enough vocabulary so that they will know how to understand your work as they continue to read your paper. Even here though, you will not be writing about your work.

The very last paragraph will be solely about your study. You must state your purpose or hypothesis. You must give the reader enough information about your methods so that they can begin to imagine how you are going to accomplish your purpose or test your hypothesis. Look at the syntax section below and choose a sentence format to start the last paragraph.

Format and Syntax

Format: When you turn in your introduction, center the word "Introduction" at the top and center of the first page. Although it is impossible to give an exact number of pages necessary for a strong introduction, you must write at least three paragraphs in order to accomplish all of the purposes of an introduction. Many particularly strong introductions, with good background research, have been about two pages.

Syntax:
The active (not passive) voice: The last paragraph of your introduction is very important. Write in the active voice when you write about what you did.

Here are some examples of sentence beginnings you can use to write that last paragraph:

I tested…   I examined….

I hypothesized that…   I compared….

I built a model to test…

The past (not the future) tense: We are taking liberties with the tense of the verbs you use in that last paragraph. When you write the introduction, your study will still be in progress. When you present your paper, in the spring, it will be finished. We are anticipating a successful completion and you are writing in the past tense, as if the work was done, so that you do not have to go back and rewrite the introduction later.

The Verb You Won't Use

~~**PROVE**~~

Prove has a very specific and formal meaning in science, and you will not be proving anything during this symposium. Don't feel badly, almost no scientist proves anything in his or her whole career. So, leave this word out of your paper.

The Difference between a Rough Draft and a First Draft

The introduction is the first section that is written for *many* people-- not just the teacher-- to read. Like the other sections, it must eventually be the *very best* writing you can possibly do. You will be required to rewrite the introduction and the other sections as often as it takes to make your writing clear for a large audience.

Never turn in a rough draft. The rough draft is your first attempt to write each section. After you write the rough draft you will need to read it, and then re write it. Repeat as many times as you can—students have told me that they rewrote sections of their papers four or five time before they turned them in.

Turn in this self revised copy as your first draft. It should be the *very best* work you are capable of. I will read and edit this first draft—and you will rewrite it and turn in the second draft.

Finally, **all** drafts are to be submitted typed and double spaced. See the Appendix.

Introduction Rubric

Category	4	3	2	1
Introduction	Inverted triangle format is followed: scientific concepts are identified and explained. Background research is thorough. Sources are cited. In the last paragraph the hypothesis or purpose of the study is stated clearly in the active voice.	Although the inverted triangle format is followed, the link to scientific concepts is not clear. Background information is not clearly explained or is not thorough enough. In the last paragraph the hypothesis or purpose is stated clearly in the active voice.	Inverted triangle format not followed. Background research is insufficient. Hypothesis or purpose is not clearly stated in the last paragraph.	Inverted triangle format not followed. No evidence of background research. The purpose or hypothesis is not stated in the last paragraph.

A Graphic Organizer for Your Introduction

Use the inverted triangle below, to plan your organization. If you wrote a strong proposal, draw freely from it!

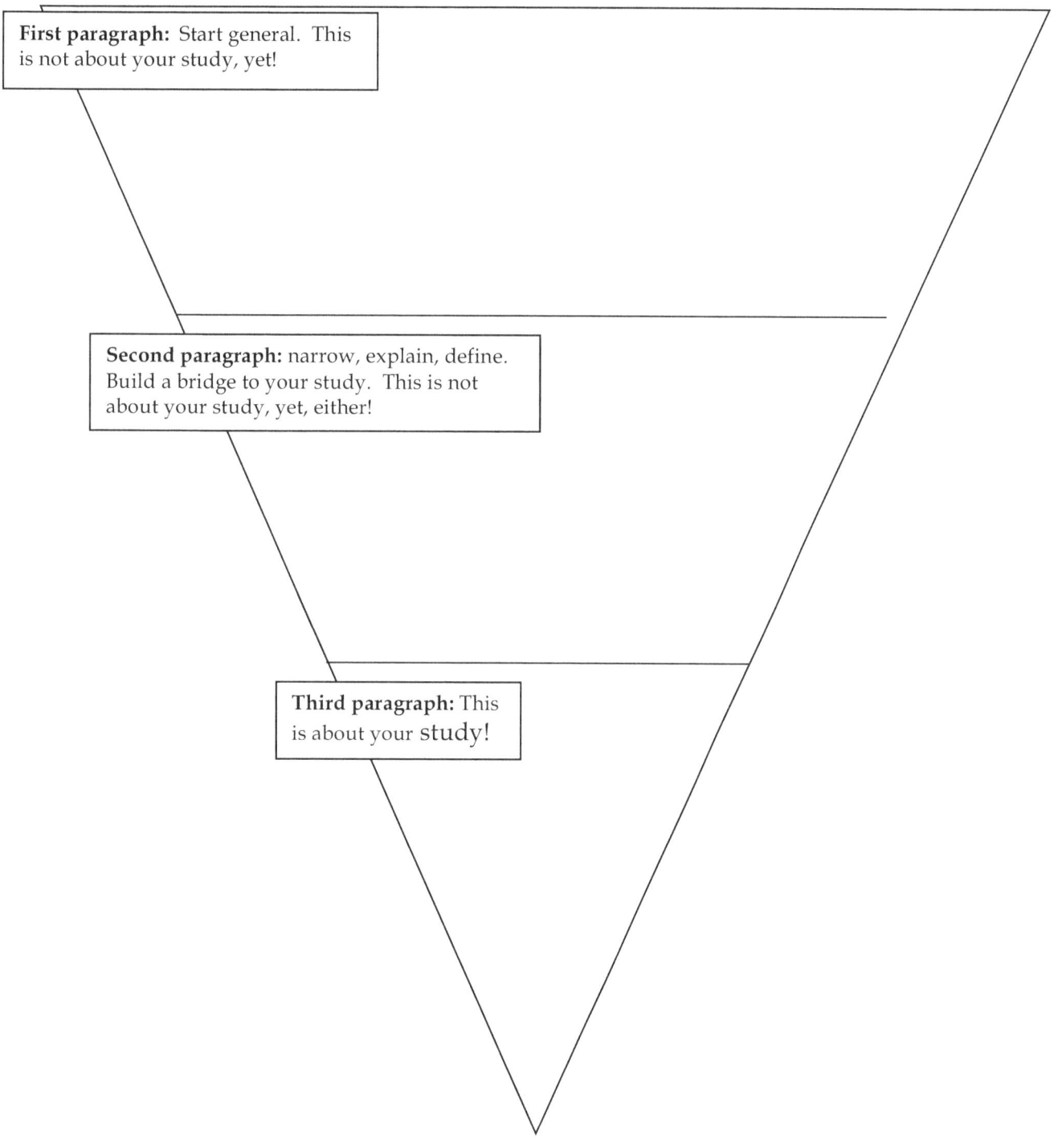

First paragraph: Start general. This is not about your study, yet!

Second paragraph: narrow, explain, define. Build a bridge to your study. This is not about your study, yet, either!

Third paragraph: This is about your study!

Checklist for the Introduction

First paragraph:
You did not write about your study at all.

Parenthetical Citations:
Look at your paper without reading it. Do you see parentheses? Inside the parentheses, are there sources cited by last name and year?

Last paragraph:
Does this paragraph contain information about your study? Is it written in the past tense?

Examples:
> If your proposal read:
> > I will test…
> Your introduction reads:
> > I tested….

> If your proposal read:
> > I will collect….
> Your introduction reads:
> > I collected….

Keeping to a Schedule
While You Do Your Experiment and Edit and Revise Your Writing

Always turn in edits as soon as you can. This is probably the most important tip I wish someone had told me.

--Ava, 7th grade

It's A LOT of work. It may seem never ending with all of the edits, green dot competitions, and even finding time to write the first draft, but by the end you will feel the accomplishment of actually going to a real symposium like some of the most successful scientists have and learn so much more than if you were sitting down reading a book. Plus you will get to create your own experiment!

-- Erica, 7th grade

Be careful when starting your project because if you hold off for too long, then if you hit some bumps in the road later on you will regret waiting until the last minute.

-- Danielle, 7th grade

Don't go home thinking that you don't have any homework. You have to keep working on your project!

-- Naseem, 7th grade

The only reason for time is so that everything doesn't happen at once.

--Albert Einstein

The science symposium is fun but can be stressful. But, once you get all your things done you feel really accomplished about what you have done.

-- Zoe, 7th grade

It's really a fun experience. It might be annoying while you're actually doing it, but you'll be glad you did it when you look back at it.

-- Kenny, 7th grade

DON'T PROCRASTINATE! Procrastinating=BAD THNGS. Procrastinating drives the quality of your work down and gives you anxiety. It is not the best solution.

--Fatima, 7th grade

Keeping to Your Schedule

As soon as your proposal is approved, you can start your study and begin collecting data. Meanwhile, you will have regular deadlines for writing each part of your paper. And, as each section is returned to you, you will have revisions to make. At any given time you could have at least three parts of the symposium to work on...and this is in addition to the rest of your school work!

The quotes on this section's title page were written in response to the questions, "What will you do differently next year?" or, "What have you learned that you will pass on as advice to a student doing a symposium project for the first time?" Two themes were repeated in the students' responses:

1. Write it well the first time—it will save you time later on.
and
2. Stay on schedule and don't miss deadlines.

Acting as your own executive

Executives are other people's bosses. They decide standards, timetables, and the general direction of projects and whole companies. Good executives are responsible. They spend time every day imagining what they want the future to be like and then working to reach that goal.

Whenever you do a long term project, you have to act as your own executive to get the work done.

You will have some deadlines to meet for class, but there will be other deadlines that you will have to set yourself.

For example, if the data are due in March, when will you collect them? How will you build in time to revise your methods if the methods of your experiment are not working the way you thought they would? The answers to these questions are unique to your project and, by acting as your own executive; you will need to make a plan to get the work done.

How?

The calendar is a reality check

The calendar is the first resource to look at when acting as your own executive. The May 6th symposium date is a firm deadline, but you will need to have all your work done well before this date. Work backwards from the data due deadline and schedule time to work on the symposium in your planner. If you need transportation you will need to schedule this with your parents, taking their schedules into account. If you are taking trips during winter and spring breaks you will need to schedule your work around being out of town.

Make good use of class time

You will have some time in class to work on your project. You can use this time to collect data, analyze data, or to write and revise. You will be evaluated on your ability to work productively during class time. Students who do not bring work to do during these periods not only waste their own time, but are distracting to their classmates. The rubric in Appendix I explains how you will be evaluated during class symposium time.

Ask for help as soon as possible

Although no one else can do your project for you, you should not let concern about the work prevent you from asking for help. If things are not going as you planned, ask me or a parent for advice or help.

Editing and Revising Your Writing

After you receive corrected work back, you need to revise it and turn it in again as soon as possible. The next class period is ideal, but I expect the work to be handed in again within a week.

Here are some editing marks that you might see in the margins or right on the words themselves:

Symbol:	Means:
ROS	Run on Sentence. You can fix this by rewriting it so that it is divided it into two or more sentences.
Frag	Sentence Fragment. These are very bad because they mean your thought was incomplete.
Capital letter with a slash through it	Make the letter lowercase.
Lowercase letter with two underscores	Make the letter uppercase.
Word circles	The word is misspelled or mistyped.
Awk.	Sentence is awkward. Rewrite it.
Arrows	Rearrange the sentences or phrases.
¶	Make a new paragraph.

Writing the Materials and Methods

Prominent Physicist Fired for Faking Data Research: Bell Labs says scientist 'recklessly' misrepresented work on microprocessors.

By CHARLES PILLER
TIMES STAFF WRITER

An influential physicist whose work in superconductivity and molecular-scale electronics seemed poised to revolutionize his field has been fired by Bell Labs for falsifying experiments over a four-year period…

"The claims were so extraordinary that if they have proved true, [a Nobel] would not have been out of the question," said Thomas N. Theis, director of physical sciences at the IBM Watson Research Center in Yorktown Heights, N.Y. **"We were all over [Schon's studies] from the very hour they appeared," he said, because the results could not be explained by known models in physics.**

"If true, they would have been revolutionary," Theis said. "We spent many months trying to replicate the results, but were unable to."…

-- excerpt from the Los Angeles Times, September 26, 2002[Emphasis added]

This is where you explain what you did. Just write everything you did exactly how it happened and you will do fine.

--Matt, 7ᵗʰ grade

…materials and methods; that's where you explain how you did your project, but don't write it like a recipe, just explain what tests you used, where you went…

--Alyssa, 8ᵗʰ grade

This is one of the easiest parts of the symposium, so don't get overwhelmed. What you can do is to make a list in your head (or on paper) of the things that you did and used, because that makes it easier for you when you actually sit down to write it.

-- Izzy, 7ᵗʰ grade

Finding good resources is really important because you want to have valid information that your can get a lot out of.

-- Dani, 7ᵗʰ grade

Write everything down or else you will have to guess amounts or go digging in the trash for old wrappers, which is hard.

--Elijah, 7ᵗʰ grade

Writing Your Materials and Method

The headline and excerpt from the Los Angeles Times, on the preceding page, reports on a particularly stunning case where a scientist faked data and was caught, in large part because no other scientist could follow his methods and get the same results.

Honesty is important in all areas of work, and scientists have several ways to insure that the work they do, and the work of their colleagues, is always honest. You have already read about how scientific journals use peer review to help insure quality publications. A second way of insuring honesty in scientific work is to carefully explain what you did, how you did the work and what you used to do the work with—so that another scientist could replicate your study.

Fraud protection is not the only reason why you report materials and methods, though. Your readers also have to know more about what you did before they can look at your data and understand your results. The materials and methods section gives the readers a description of your study that will allow them to picture your work. By writing a clear description, you will have an easier time later when you teach your readers about the important parts of your data.

Format and Syntax

This section is always written in paragraphs, never in lists. You will use the past tense because it is about work you did. (Just like with your introduction though, you may be in the middle of collecting data. So we will anticipate a successful completion of your project and write in the past tense now, to save time later.)

You will have to decide how to organize the paragraphs so that they make sense to the reader. This almost always means that you will not be describing the history of what you did-- but rather, grouping the categories of what you did. Your materials and methods section will not read like a diary.

For example: Imagine that you sampled water from a nearby stream and tested it for pH throughout the winter. In your materials and methods section you would group the categories of activities. The topic sentences of each paragraph might read like this:

I collected water from _____stream ten times throughout January and February.

To test the pH I used a pH pen from Acme Science Co. (Part #0000).

I analyzed and graphed the data using Microsoft Excel (Microsoft 2002).

If you do an outdoor study: My study site was located _____ in Dutchess County, NY.

Each example sentence given above would begin a paragraph that has more explanation about that part of your materials and methods.

Make sure you include (in the order that makes sense for your work)

What you sampled (or measured).
When you sampled (or measured).
Where you sampled (or measured).

If you have a field site for your study, use Google Earth to find the longitude and latitude of the site and include this information in your site description.

The manufacturer and product number of anything you bought or used.
Chemicals you used.

Standard lab equipment does not need to be described. For example, if you used vinegar and water to make a pH solution, you do not need to explain that you used a beaker and a graduated cylinder.

You do not need to give directions for using equipment that comes with instructions. You should write that you followed the manufacturer's directions—unless you didn't, in which case you need to add how you varied from the directions.

When giving measurements, state which units you used. Did you measure to the nearest millimeter? Whole degree Celsius? Milliliter? (Note: This is your reminder that ALL measurements MUST be in metric!)

When you write about collecting data or samples, use the verb *collect*—don't write that you "took" the data, or that samples were "taken." That makes you sound like a thief!

You do not need to include any sentences about the ultimate purpose of your study. You audience read the purpose in the last paragraph of your introduction—just before they turned the page for the materials and methods section!

Finally, the good news…

Many students find this to be a straightforward and relatively quick section to write!

Materials and Methods Rubric				
Category	4	3	2	1
Materials and Methods	Materials and methods were explained with enough clarity that a reader could replicate the study. Flow of paragraphs was organized by category and not like a diary. Equipment and parts were specified appropriately.	Materials and methods were not explained clearly enough to be replicated. Flow of paragraphs was not organized. Some but not all equipment was described correctly.	Materials and methods had gaps that made the reader unable to understand the study. Flow of paragraphs was not organized. Equipment was not described correctly.	Materials and methods were listed, and not written in paragraph form.

A Sample Materials and Methods

I collected gray water runoff from an outdoor pipe at my home. The gray water was from my bathroom sink, shower and bath. Grew Bush Beans (Burpee Seed Co.) in individual 5 cm peat pots, with one bean planted per pot. I planted one bean seed in each pot. The control group had twelve plants and the gray water treatment group had twelve plants. Every other day, I watered them with sixty milliliters of water. Every day I rotated the plants so they got exposure to sunlight. Every other day for one month, I measured every plant and counted how many plants there were. I measured in centimeters. I grew the plants from January 29th to March 5th, 2011. I started

measuring them on February 5th.I analyzed and graphed the data using Microsoft Excel 2008.

Notice that someone else could use this materials and methods and do a similar study. They would know which brand of bean to buy and how to treat the plants. They would know what to measure and how often. They would know which program did the analysis.

This section doesn't have to be very long, but it has to have the correct details in it.

Checklist for your Material and Methods

Did you include numbers?
Look at your paper without reading it. Most Materials and Methods sections have numerals in them, not just words. This is because:

> Product numbers are included
> and/or
> Longitude and latitude are included

Did you include these?

Did you write in organized paragraphs?
Look at your paper without reading it. Materials and Methods sections are written in prose—all sentences—not using lists.

Is your paper all paragraphs without lists?

Does your paper have only metric measurements?
(No inches, ounces, cups, degrees F…)

Is it written in the past tense?

If you are using Microsoft Excel to analyze your data, include this in your Materials and Methods section.

Writing the Results

Design is the conscious and intuitive effort to impose meaningful order.

-- Victor Papanek, architect and designer

This part is simple, but it must be done with the figures in front of you. All you have to do is find the trend and state it in your results. There might be more than one trend for each.

--Sarah, 7th grade

Be sure to relate to your figures, and get to the point. Include important things, and things that would really describe your results.

--Adam 7th grade

When writing your results, if you study your data well, the writing comes a lot easier than you would think.

--- Avery, 8th grade

This part should be pretty easy. All you have to do is write what you saw in your experiment. You have lived and worked on your experiment for so long that you should just know what to say.

--Elizabeth, 7th grade

When making the figures on the computer, it helps a lot if you have drawn it out on paper first. Think of math – what are the independent and dependent variables?

--Sophia 7th grade

The results section is really easy to write. It is really important to remember that you are only stating what happened, not why.

--Sophie, 7th grade

You do not need to be elaborate because you're just telling what happened.

--Patrick, 7th grade

A man should look for what is, and not for what he thinks should be.

-- Albert Einstein

Writing Your Results

Before you can write your results, your study must be finished. All of your data must have been collected. At this stage, it is called *raw* data. Your job is to understand it, and *design*, in prose (i.e. sentences, paragraphs) and figures and tables, the best way to make it understandable to your audience. No one else will see your raw data—you must present data that has been analyzed.

Exactly how you will present your data depends on the nature and results of your study. Because you were required to collect quantitative (numerical) data, everyone will have at least one graph or table. Many studies will have one or more of each.

How do I begin?

This section is entitled writing your results, but no scientist can start by *writing* about his or her data. Everyone must start by studying and thinking about his or her data.

Your first step is to look at the raw data. Decide how it will be analyzed. Should you make a graph? Calculate means? For most students, this is not a difficult decision—when you designed the study you had some idea of what the categories of data would be.

Do the analysis. For example, if your data are showing a change over time, make a graph. If you are comparing the measurements of more than one group, calculate means.

Sit and think. **"Talk to the wall."** What do your data show? What are the trends? Now you can place your data in tables and graphs that you design to best show the trends you notice.

Some rules

<u>You will only have figures and/or tables</u>
This means nothing will be called a chart, graph, picture, diagram, drawing…or anything else, in your results or anywhere else in your paper. Does this mean that you can't have charts, graphs, pictures, diagrams, drawings, or anything else in your paper? NO! The words "table" and "figure," are simply how you will refer to everything that is not prose.

If you present the data in a table, refer to it as a table. Everything else-- photos, graphs, drawings-- are referred to as figures. [I know you want to know why. I can't "figure" it out either-- except that this makes the system of referring to them easier for the writers.]

<u>You will point out the trends but not explain why</u>
For example: The average temperature of the pond water was…The average temperature of the stream was…

Do not write: The pond was colder because it was at lower elevation.

Does this mean you will never explain anything?

No! You will explain EVERYTHING in the discussion section.

<u>Write about every single figure or table</u>
Never add a figure or table without writing something about it. You will guide your readers when to look at the tables and figures in the prose (sentences and paragraphs) of your results section. (You must follow this rule in other sections too. If you have a figure that is a photo of your research site and you want the readers to look at it when they read your materials and methods, you refer to that figure in the materials and methods section.)

Here are some sample phrases

Figure 1 shows a diagram of the watershed model. Water was added at label "A" and recovered at label "B."

Table 1 shows a comparison of mean plant growth. The plants treated with fertilizer grew approximately twice as high as the plants without fertilizer.

As can be seen in Figure 2, conductivity increases with sodium concentration.

The plants watered with acid did not grow as large as the plants watered with distilled water (Figure 3.)

Notice that in each of these phrases, the writer has described the trend in the data — and directed the reader to look for this also.

Format and Syntax

For tables and figures: These must be numbered and referred to in the order in which they appear. Your first table is Table 1; your second is Table 2, etc. Your first figure is Figure 1; your second is Figure 2, etc. As you revise your sections you may need to renumber.

When you refer to the figures and tables, you are referring to them by name, therefore the words figure and table are capitalized, as in "As can be seen in Figure 1…." Because it is a title you can use the numeral rather than writing out the number of each figure or table. For example write "Table **1** shows," not "Table **one** shows…"

The tables need titles and these are placed at the top of the table. You can easily make tables in Microsoft Word.

Figures need titles and these are conventionally placed at the bottom of the figure. If your figure is not a graph, print a title and place it under the figure.

Titles are descriptive and are written like phrases and sentences, not like book titles.

For figures that are graphs, you also need to label the X and Y axes! The labels must include the units of measurement (in metric of course) that you used.

Print all figures that are graphs out on a single sheet of paper—make them large!

Syntax: When writing the prose of your results you will use both past and present tense. The results of the work will be in the past tense (e.g. "The plants watered with salt solution died."). When referring to what the tables and figures show, use the present tense (e.g. "As Figure 1 shows..."), because the figure or table will continue to show this trend as long as the paper exists!

Two Points of Caution

Do not be surprised if you spend more time studying your data and thinking than writing. If you do the thinking part well, the results will practically write themselves.

and

Do not confuse "design" with "decorate." Excel gives you many fancy choices for graphs—and almost all of them are clunky and will not make your data clearer.

Results Rubric				
Category	**4**	**3**	**2**	**1**
Results	Trends in the data were summarized in prose. The figures and/or tables were clear and labeled and referred to correctly.	Not all trends in the data were summarized. The tables and/or figures were not labeled or referred to correctly.	Trends in the data were not summarized. Data were not presented in tables and/or figures.	Writing does not indicate that a study was done, or that the data were understood by the student. Data were not presented.

Checklist for your Results section

Editing figures takes a keen eye.

Are the x and y axes labeled?
Is the figure titled?
Does the title include its number?
Does the figure make sense?

Editing tables takes at least a little sense of style.

Are the rows and/or columns titled using a bigger or bolder font?
Are the cell sizes appropriate?
Is the table titled?
Does the title include its number?

Did you include prose? Did you write about the trends shown in every figure and/or table?

Writing the Discussion

This is probably the most complicated part of the whole symposium because this is the part where you answer questions that haven't been asked yet.

--Gabrielle, 7th grade

Connecting your project to the world will make you understand and appreciate your project better.

--Maddy, 8th grade

You have to find out why things happened like they did.

--Emily, 8th grade

Eveything has a why....you just have to find it.

--Noah, 7th grade

The discussion is the hardest section, but don't get nervous. If you research, and really understand your project, you'll be fine. If you have any questions ASK LAURA don't just wing it because then its harder to do the edits.

--Lulu, 7th grade

When writing your discussion you want the Science Symposium to be over with already. Do not think that way! You want to think about the discussion as the last mile of the race. You need to catch up and put all your effort into it. You want to think hard about everything you did with your project and explain it to the world! The discussion is probably the most important part of your study.

--Julia, 7th grade

This is a hard one. It could take as many as four drafts to get the Discussion perfect. Be prepared to have to go over it, and over it... a good way to get it done is to write a very rough draft, hardly more than notes, and then go over it and correct it as you edit.

--Maddy, 7th grade

If we knew what it was we were doing, it would not be called research, would it?

--Albert Einstein

Writing Your Discussion

This is it…your last major section. Here is where you explain everything to your audience. How do you explain your data and draw a scientific conclusion?

The spotlight starts on you, but moves to the rest of the world and the future

Just as your introduction was a funnel that started wide and ended narrowly, with the purpose of your study, the discussion section is also a funnel—only going from the narrow (your data), to broad (the rest of the world and the future).

You have four purposes in the discussion section:

1. Clearly link the trends you pointed out in your results section, to the purpose or hypothesis in your introduction.
2. Explain the science behind why you got the trends in data you described in the Results section.
3. Draw a conclusion about what this means for the rest of the world (and humanity if appropriate).
4. Suggest areas for future study based on your findings.

Where do I begin?

Begin by thinking, not writing. Most students have a mystery about the trend their data showed. The mystery is what you want to explain. Identify your mystery and make it the theme that runs through the discussion section.

When you start to write, you can assume that your audience has read the results section. You do not need to repeat every statement you wrote in that section.

Because most of your data are in well designed tables and figures, one way to start is by explaining why the data look as they do in those tables and figures. Write as many paragraphs as you need to explain why you got the results that you did.

Here are some examples of topic sentences that might begin paragraphs in a discussion section:

> The plants in the _____treatment grew higher *because* they received more_____.

Or

> My study shows that temperature has a big effect on _____*because* I saw_____.

Notice that these sentences have built in "memory refreshers" for the readers because words in the sentence refer directly to the treatments and the trends in the data.

Relate your discussion back to your hypothesis or purpose

Each paragraph that explains your data will connect to your hypothesis in some way. Either your data support your hypothesis or they do not. You will state this explicitly. If your study had a purpose you must reach a conclusion about what the data mean for the purpose. This is the start of using both your data and your understanding of it to draw a conclusion.

Sometimes data are not all what you predicted. You will not write that your hypothesis was wrong. If your data do not support your hypothesis, then state, "My data did not support my hypothesis." This does not mean that you did not practice good science. We learn as much from data that are not as predicted as we do from data that are as predicted. Isaac Asimov wrote that the most exciting words in science were not "Eureka!" but, "That's funny." This is the reaction of a mind that is curious and about to look into the observation further. This is how new discoveries are made.

Here are some sentence structures you can use to get started:

I predicted _____ and the results showed _____.

The data supported the hypothesis that _____.

The data did not support the hypothesis that _____.

This was because_____

That part about the rest of the world (and humanity if appropriate) might have you worried

Remember that right from the beginning you were relating your work to the rest of the world. This is why your study was relevant and interesting. Of course you did not consider the world as a whole, but focused on some important issue. This was how you started your introduction, and now you must draw a conclusion or two that relates your data to that part of the world.

Sometimes students do not know why they got the results they did. Perhaps the data were surprising. In this case, you have to learn more! Your data are not wrong-- you followed a procedure and got results! If the data do not seem compatible with your expectations, you still have to explain them. You are in the best position to do this-- the data are yours!

Spotlight on the future

If you were going to turn this work into a bigger study, or if you were going to turn this work over to another student to pursue in more depth, what would the follow up studies be? This is your legacy-- an ability to take what you learned and think about the future. How can what you learned become the basis for more work? Make predictions or hypotheses about what the data of a follow up study would mean.

Remember
Do not use the word "PROVE" anywhere in your paper.
Appropriate substitutions include "shows" and "demonstrates," or "supports."

Discussion Rubric				
Category	**4**	**3**	**2**	**1**
Discussion	All four purposes are met: The results are explained in a way that relates to science and the purpose or hypothesis of the study. The conclusions are related to the rest of the world. Strong suggestions are made for future research.	Although a conclusion is drawn, it is not well explained in terms of science. The conclusions are weakly related to the rest of the world. Suggestions for further research are not feasible or not related to the study.	Conclusions are not supported by the data. Conclusions are not related to the rest of the world. No suggestions for future research are made.	No conclusion was apparent OR important details were overlooked.

Writing the First and Last Words they Read

Title Page

Acknowledgements

Literature Cited

Only give people credit where credit is due, If you looked at a false website then don't give it credit.

"Affects" and "Effects" mean different things.

Acknowledge everyone involved.

--Tyler, 7th grade

Bibme.com really helps put together your Literature Cited properly. There is nothing worse than having everything done but your Literature Cited is in the wrong format.

-- Ella, 7th grade

Record every source as soon as you decide to use it. Keep a document that is only Literature Cited, and then you've got your section done ahead of time.

--Eve 7th grade

It is fun to make the acknowledgements humorous.

--Elinor, 7th grade

Bibme.com is a good place to start your literature cited but there bib-me does not do everything; you still have to do some things by yourself.

--Alex, 7th grade

Use alphabetical order and hanging indent.

--Josh, 7th grade

Writing your Title, Acknowledgements and Literature Cited Sections

The first thing they read is the title...

Sometime between writing the proposal and the introduction, you will compose a working title. You can change this anytime before the program for the symposium itself is written.

Science titles are declarative and informative. They are not like book titles. You can look at titles from symposium programs of prior years to get some ideas.

Thanking you

Even though you are the researcher, you did not work alone. Others helped you by buying materials or driving you to study sites. Write a short paragraph acknowledging people who helped you, and thanking them.

Literature Cited

Remind everyone to write down what date they used the web sites....
--student suggestion

This is the very last section of your paper. You have been working on it all along. All sources that were cited in your previous sections will be listed here. If a source is not cited in your paper, do not include it in your literature cited section.

Congratulations! Your writing is done. Now you just have to put the poster together, practice your talk and you are ready to go!

Designing the Poster

It is really fun making your poster. Make it you but do not overcrowd your poster. You want your poster to look professional and get the point of your project across. You want presentation more than creation. Think about being able to read it, before you add the glitter glue and extra "add- ons"

-- Julia, 7ᵗʰ grade

Make the font big and make your paper cuts even.

-- Peter, 7ᵗʰ grade

This is the best part, so make sure that you have all of your work done so that you don't have to be stuck editing work when everyone else is decorating their poster.

-- Ellie, 7ᵗʰ grade

Be creative. The one key tactic that makes this job easier is getting all your green dots a week before the science symposium. You have a lot more time to be creative and artistic with your poster if you're not rushing it a day before the science symposium.

--Mac, 7ᵗʰ grade

Making the poster is the most fun thing to do in the science symposium. With your poster relax and have fun doing it. You can do whatever you want, but you want a board that stands out. Try to get all your green dots quickly so you can start working on your poster sooner.

--Zoe, 7ᵗʰ grade

Making your poster is a lot fun you don't want to be the kid who is sitting in the classroom while everybody else is working on there poster.

--Rita, 7ᵗʰ grade

This was really fun and relaxing after all the work.

-- Brooke, 7ᵗʰ grade

Don't make your poster too crowded, so people can understand it better, take it in in stead of getting lost in the design. The best part of the poster is the science of the poster.

--Jenna, 7ᵗʰ grade

Designing your Poster

Most students consider this to be fun and rewarding after all the work of the previous months. You have flexibility in designing your layout, with a few important format guidelines.

Format

You will be given a display board that stands alone and is divided into three sections. They are all white.

You must have the title of your paper at the top of the center section. Your name must be near the title.

You must print out each section and mount it on colored backing paper before gluing it to the board. I have background paper, but you are welcome to buy your own if you want something special.

You must use a clear font and you must use as LARGE a font point size as you possibly can. When you wrote your sections they were double spaced in 12 point. Twelve point is too small to be read at a distance. Make a new file of each section and experiment with larger font sizes.

Keep the classroom work area neat. Do not waste resources like paper and glue.

Talking about your Project to Visitors at the Symposium

When you talk, include your subject, your hypothesis, what actually happened, and why. Then you can show your figures, and explain what they mean. Figure out a way to end what you say, so you don't end up trailing off at the end, and feeling dumb.

--Talia, 7th grade

Being at the Science Symposium is nerve racking at first but as you explain it will get easier. Each time you explain it you end up adding something to your main explanation, as people point out could suggestions to you.

-- Maggie, 7th grade

For being at the symposium just give people a whole speech and answer their questions and be patient with them.

--Lily, 7th grade

Come up with a pattern for what you are going to say – something like 1) Introduce your project, 2) Explain why you chose it and how it is relevant, 3) What happened, 4) Support your theory and show them your tables and figures, and 5) Conclude!

-- Kelly 7th grade

Stand tall and look the person in the eye.

--Josh, 7th grade

It was really busy and crowded it was basically like every family member or friend was there to see what we came up with and how we completed it all.

--Reilly, 7th grade

Speak when spoken to. Answer the questions asked and refer to your graphs often. The most common questions people will ask are "What's your project" and "What did you find out." Eventually you'll get into the habit of having a set speech. Keep water and food with you. You will get thirsty and hungry.

--Jesse, 7th grade

If you can't explain it simply, you don't understand it well enough"

--Albert Einstein

Talking about your Project

All the advice on the cover page for this section is worthy and you should read it carefully.

What you can expect

The dining room will be full (at least 150 people, more like 200 will be passing through.) Therefore, it will be noisy.

You will be repeating what you say for about an hour and a half as new people visit your poster.

Most visitors will want to talk to you rather than just read the poster. They may just start with the basic question "What was your project about?"

Most visitors will ask you questions for clarification and you will not find them difficult to answer.

How to take command of the situation and show what you know

Smile and say hello.

Make sure to make eye contact with your visitors. This will help them to hear you better. It is also polite, conveys confidence, and indicates that you are interested in talking to them.

You can and should point to your data when you are speaking about it, but turn back and face away from the poster at other times.

If you are asked a question that you don't immediately know the answer to, don't panic! You do not have to know the answer to all questions. Next, stop and think. Try to repeat the question aloud-- if you do not know the answer you can still say something intelligent about what you think the answer might be. Finally, compliment the visitor on asking a good question-- they might have an idea about the answer themselves that they want to share with you.

Make physical or mental notes of any insights you get as you present your work, answer questions, or listen to visitor's comments. You will have the opportunity to make small changes to your final paper based on these insights.

Above all, realize and remember that you know more about your project than anyone else in the room!

Appendix

Formatting

Requirements and Hints

Examples of Literature Cited Entries

Formatting your Paper

Every draft that you turn in must be:

Typed, in 12 point font

Double spaced

Have right and left margins of 1inch each

Have the title of the section (e.g. Introduction, Materials and Methods, Results, Discussion, Acknowledgements, Literature Cited) centered at the top of the first page

Have your name at the top of the first page.

Be stapled if the writing is more than one page long.

Electronic copies

Always turn in a hard copy.

Always have an electronic copy AT SCHOOL.

Methods to have an electronic copy at school in order of reliability:

> laptop backed up
> Google docs
> e-mail your document to yourself at school
> use a memory key
> burn to a CD

Do not turn in electronic copies--turn in only hard copies. This means that in order to meet the deadline, you must print out your paper BEFORE CLASS.

Words to avoid

These words should be avoided in your writing:

> **thing:**   Replace with the actual *name* of the…thing.
>
> **then:**   Leave the word out and reread the sentence. If the sentence still makes sense, you should omit the word.
>
> **they:**   When referring to a treatment group. *Name* the treatment for the readers-- it is too confusing for the readers to remember all the treatment groups.

you: "You" will not be writing in second person. The chatty style is not appropriate for this formal writing…"you" can be chatty at the symposium…

prove/correct: Reread the *Writing the Introduction* section for why, and the *Writing the Discussion* section for alternatives.

Finally…

Spell check, but also READ your sections before turning them in.

Keep hard copies of older drafts.

Turn in first drafts, not rough drafts.

Make revisions and resubmit as soon as possible, but not more than one week after getting a version back.

Format for Citing Sources

We are using the Modern Languages Association (MLA) format, or the format you are being taught in your humanities class.

Here are some examples of citations from sources that you will probably use (these are all fictitious). There are more examples posted in the classroom. If you have questions, ask!

Example of a book citation:

Smith, Thomas. How to Succeed at your Science Symposium, New York: Imagination Press, 2005.

Note: If the book has more than one author, only the first author's name is listed with the last name first. The other authors are listed first name first (e.g. Smith, Thomas and Jane Ford).

Example of a newspaper citation:

Wilson. Rita. "Students Succeed at Symposium." Poughkeepsie Weekly 5 May 2006: A1.

Example of a web site citation:

Smith, Thomas. Succeeding at your Symposium. 5 January 2004. Accessed 8 August 2005 <http://www.thomassmith.com/sciencesymposium/>.

Note: The first date is the date the page was last updated or posted. The second date is the date you accessed the site and took information from it.

Also, notice the **hanging indent**. The way to set hanging indents depends on your computer's operating system and software.

Due Dates for 2012 Symposium

Section	Due WEEK of:
Research Questions	December 12
Research Proposal w/ Literature Cited	January 23
Begin actual project	January 23
Introduction w/ Literature Cited	February 13
Materials and Methods	March 5
Data (research completed)	March 12
Results	March 19
Discussion w/ Literature Cited	April 16
Poster	April 30
Symposium	Friday, May 4

The *specific* day of the week that each section is due depends on your particular class schedule. The date will be announced and posted in class, and on the course web site and course calendar. Please add these dates to your planner or task list.

Planned absences are *not* an acceptable excuse for a missed deadline. If suddenly ill on the day a section is due, e-mail it and bring a copy in on the first day you come back to school—even if you don't have class that day.